A Magnet's Strength

Magnets attract items that are made of iron and steel. Most magnets are made of iron. A magnet's force can pass through air, water, and even solid objects!

This magnet is pulling many paper clips. Can you count them?

Magnets have many uses.
Magnets are used in many places.

Some magnets must be very strong to do work. This magnet is used in a scrap metal yard. How many pipes is the magnet attracting?

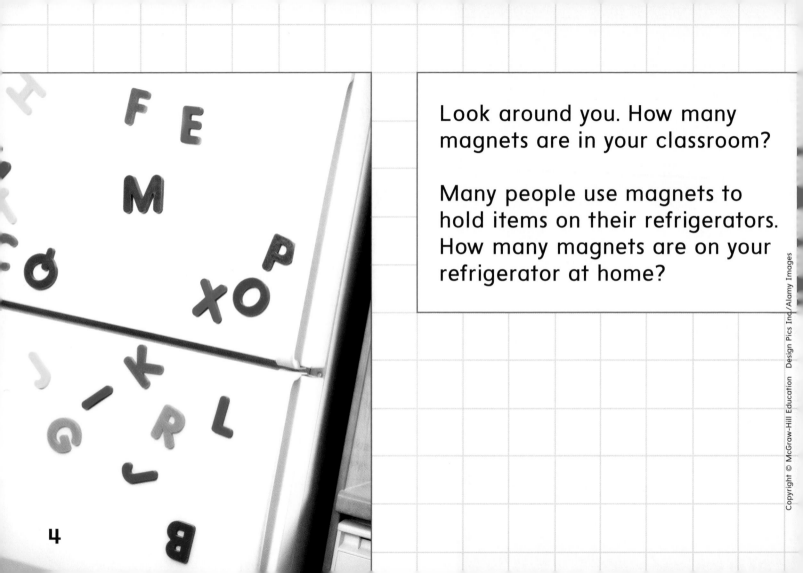

Look around you. How many magnets are in your classroom?

Many people use magnets to hold items on their refrigerators. How many magnets are on your refrigerator at home?

4

You can experiment to see if something is magnetic.

First, place a magnet by an object. Does the magnet pull or push it? If the answer is yes, then it is magnetic.

You could also see whether an object sticks to a magnet. If it does, then the object is magnetic.

Which of these objects do you think are magnetic? Which ones are not?

Paper clips

Cereal

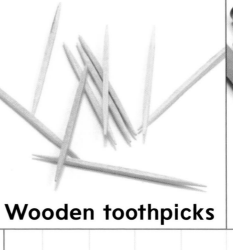

Wooden toothpicks

Crayons

Try the tests that you did on page 5. Which of the items on this page are magnetic? What do the magnetic items have in common?

Plastic buttons

Screws

Pins

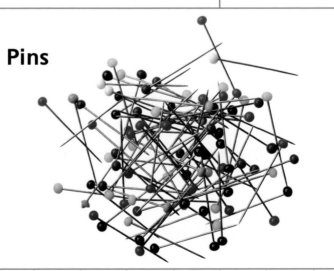

Here are three types of magnets. How can you tell which magnet is the strongest?

Disk magnet

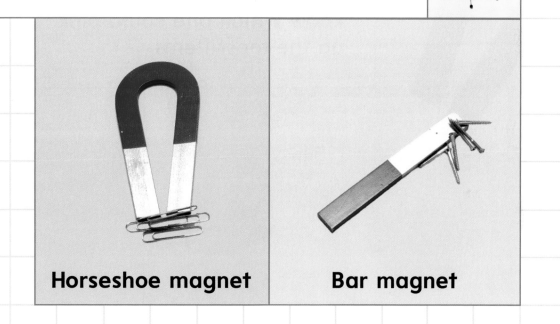

Horseshoe magnet

Bar magnet

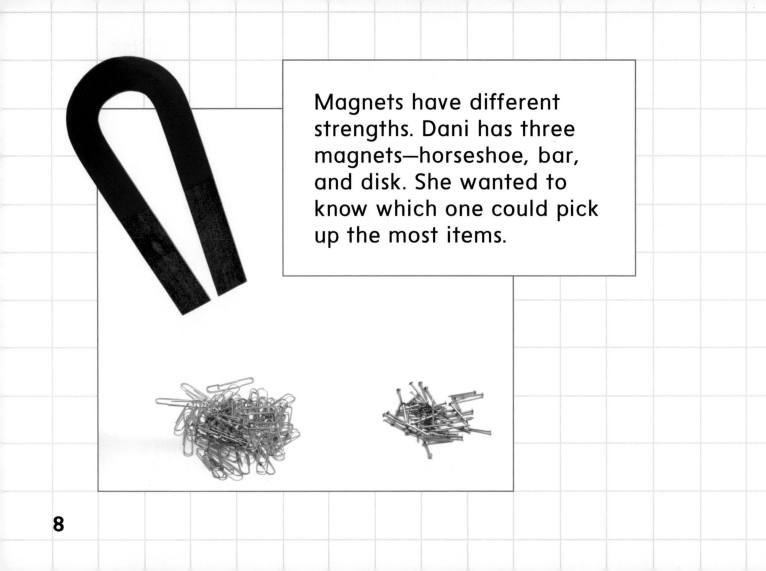

Magnets have different strengths. Dani has three magnets—horseshoe, bar, and disk. She wanted to know which one could pick up the most items.

Dani tested the horseshoe magnet first. She placed the magnet near paper clips. Then Dani placed it near small nails. She added the number of nails and paper clips the magnet attracted. In all, Dani counted 99 paper clips and small nails.

	50		
	49		
Total	99		

Dani then tested the bar magnet and disk magnet. Her results are below.

The bar magnet pulled 79 items. The disk magnet pulled 33 items.

	50	38	22
	49	41	11
Total	99	79	33

Dani displayed the data in
two bar graphs.
The graphs are shown below.

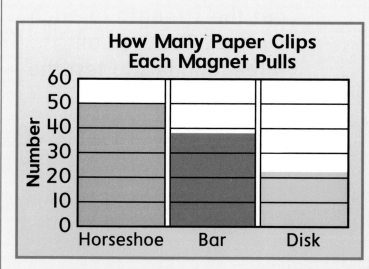

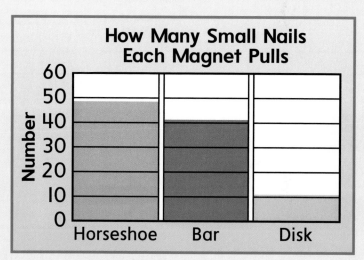

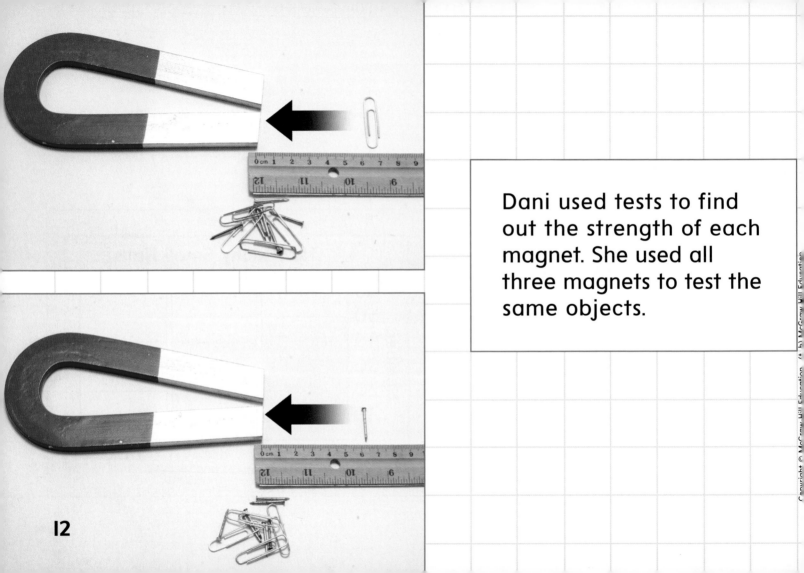

Dani used tests to find out the strength of each magnet. She used all three magnets to test the same objects.

Dani tested the horseshoe magnet
with a nail and a paper clip.
The horseshoe magnet pulled
a nail from a distance of 6
centimeters.
It pulled a paper clip from a
distance of 5 centimeters.

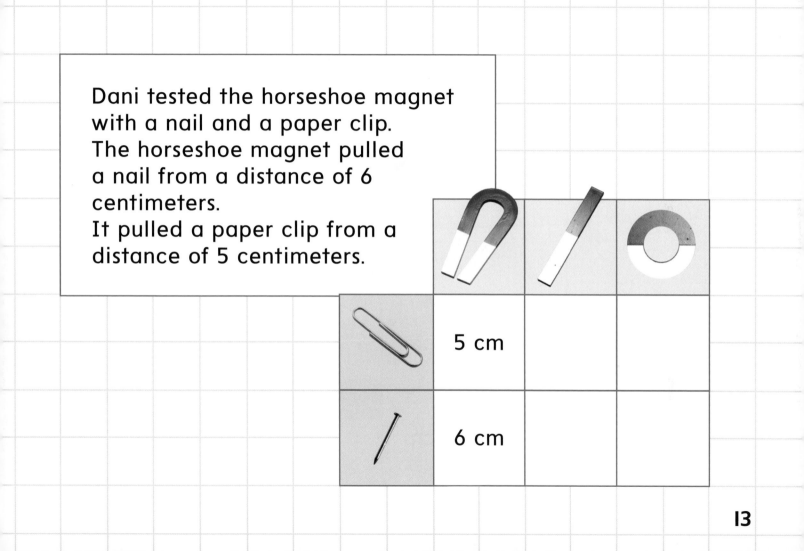

	5 cm		
	6 cm		

Next, Dani tested the bar magnet and the disk magnet.

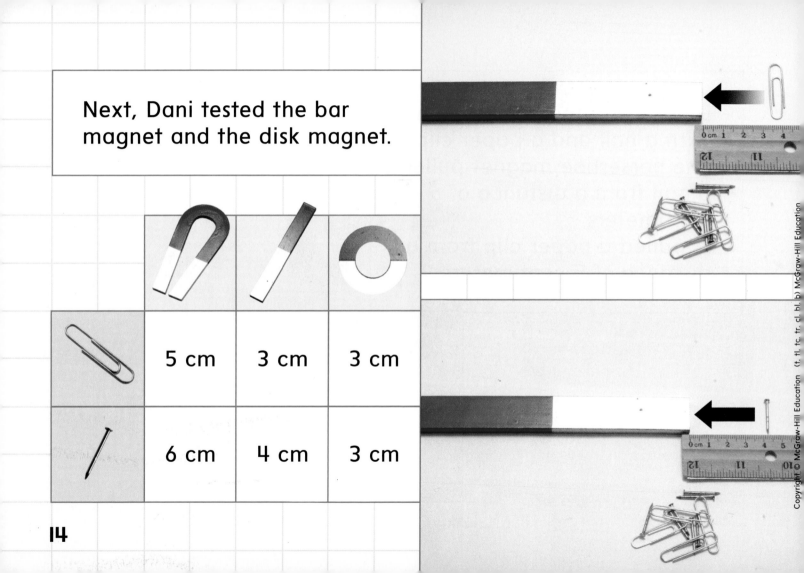		
5 cm	3 cm	3 cm
6 cm	4 cm	3 cm

14

Dani displayed her data in two new bar graphs.

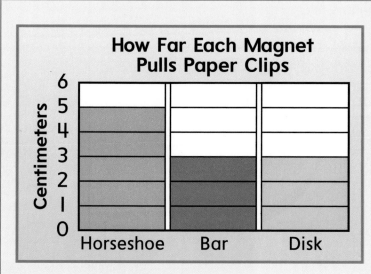

How Far Each Magnet Pulls Paper Clips

Centimeters

Horseshoe   Bar   Disk

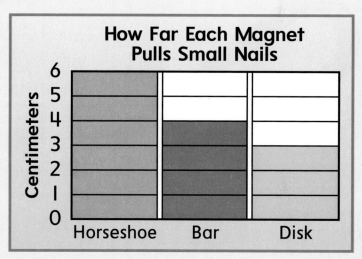

How Far Each Magnet Pulls Small Nails

Centimeters

Horseshoe   Bar   Disk

Study the two sets of bar graphs on pages 11 and 15.
Based on Dani's tests, which magnet is the strongest?

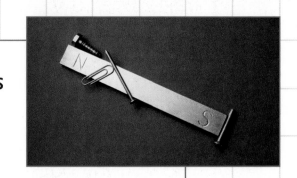

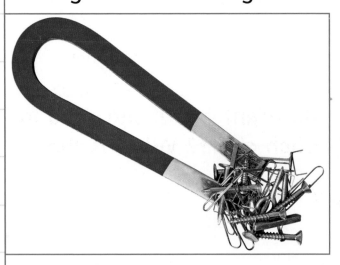